BRITISH RAILWAYS DIESEL ELECTRIC
CLASSES 44 TO 46
The Mighty Peaks of the Midland Main Line

Front Cover: 46032 (ex D169) passes Belle Isle on 22 March 1980 while working a Kings Cross–Bounds Green stock move.

Rear Cover: Class 45/1 45125 (ex D123) stables in Bedford Carriage Sidings on 13 March 1983 during a period when a shortage of Class 127 DMU trainsets resulted in the operation of locomotive + coaches on some Bedford–St Pancras suburban services.

BRITISH RAILWAYS DIESEL ELECTRIC
CLASSES 44 TO 46
The Mighty Peaks of the Midland Main Line

FRED KERR

AN IMPRINT OF PEN & SWORD BOOKS LTD.
YORKSHIRE – PHILADELPHIA

First published in Great Britain in 2022 by
Pen and Sword Transport
An imprint of
Pen & Sword Books Ltd.
Yorkshire - Philadelphia

Copyright © Fred Kerr, 2022

ISBN 978 1 39908 994 4

The right of Fred Kerr to be identified as author of this work has been asserted by him in accordance with the
Copyright, Designs and Patents Act 1988.

A CIP catalogue record for this book is available from the British Library.

All rights reserved. No part of this book may be reproduced or transmitted in any form or by any means, electronic or
mechanical including photocopying, recording or by any information storage and retrieval system, without permission
from the Publisher in writing.

Typeset in 10/12 Palatino by SJmagic DESIGN SERVICES, India.

Printed and bound by Printworks Global Ltd, London/Hong Kong.

Pen & Sword Books Ltd incorporates the imprints of Pen & Sword Books Archaeology, Atlas, Aviation, Battleground,
Discovery, Family History, History, Maritime, Military, Naval, Politics, Railways, Select, Transport, True Crime, Fiction,
Frontline Books, Leo Cooper, Praetorian Press, Seaforth Publishing, Wharncliffe and White Owl.

For a complete list of Pen & Sword titles please contact

PEN & SWORD BOOKS LIMITED
47 Church Street, Barnsley, South Yorkshire, S70 2AS, England
E-mail: enquiries@pen-and-sword.co.uk
Website: www.pen-and-sword.co.uk

or

PEN AND SWORD BOOKS
1950 Lawrence Rd, Havertown, PA 19083, USA
E-mail: Uspen-and-sword@casematepublishers.com
Website: www.penandswordbooks.com

CONTENTS

British Railways Diesel Electric Locomotives: Classes 44 to 46   6

Section 1   Class 44: D1 - D10 / 44001 - 44010   8

Section 2   Class 45: D11 - 137 / 45001 - 45077; 45101 - 45150   18

Section 3   Class 46: D138 - D193 / 46001 - 46056   91

Section 4   Headcode Boxes   112

BRITISH RAILWAYS DIESEL ELECTRIC LOCOMOTIVES: CLASSES 44 TO 46
by Fred Kerr

When British Railways (BR) inaugurated its Pilot Scheme as an early stage of the 1955 Modernisation Plan, Derby Works looked to offer a design based on the successful Derby Twins (10000/1) which it had produced in 1947/48 as the London Midland Scottish Railway's contribution to the nascent BR. This pair of locomotives had been developed in conjunction with English Electric who had supplied the engines but, for the Pilot Scheme, English Electric had decided to provide its own design for contention hence Derby Works was forced to look elsewhere for its engine. Its search led to the firm of Sulzer Brothers, a Swiss engine builder which not only had the single most powerful engine of the day rated at 2300 hp @ 750 rpm but had also developed its engine technology within the UK in cooperation with the firm of Crompton Parkinson who supplied electrical traction equipment. This allowed Sulzer Brothers to provide a complete drive chain for the proposed design.

The Pilot Scheme had identified three power ranges for consideration; Type A (rated at 800-1000 hp), Type B (rated at 1000-1500 hp) and Type C (rated at 2000+ hp) which later became Types 1, 2, and 4 respectively. As work began on Derby's Type C design, the civil engineer expressed concern that the new locomotive would breach the 20-ton axle weight that was currently in vogue, especially given that the projected weight of the design was expected to be 138 tons, hence his insistence that the bogie have four axles, which included a non-powered pony wheel. The only suitable bogie design at that time was the 4-axle bogie fitted to the Southern Railway's (later BR (Southern Region)) prototype design (10201-03) and this was imposed on both the Derby Works and English Electric designs.

Derby Works supplied designs to both the Type B and Type C specifications using Sulzer engines, but its Type B design (later BR Class 24) D5000-19 used the 6LDA28 engine rated at 1160 hp @ 750 rpm combined with AEI electrical equipment as the competing Birmingham Railway Carriage & Wagon Company offering (later to become BR Class 26) had adopted the Sulzer/Crompton Parkinson drive chain while its Type C design (later to become BR Class 44) was able to adopt the Sulzer/Crompton Parkinson drive chain that combined the Sulzer 12LDA28 engine rated at 2300 hp @ 750 rpm with Crompton Parkinson electrical equipment. The 12LDA28 engines for the first ten locomotives (later Class 44) were supplied direct from Sulzer's Swiss factory at Winterthur but subsequent deliveries for both Class 45 and Class 46 orders came from Vickers at Barrow in Furness as sub-contractors to Sulzer Brothers to produce the LDA28 series of engines.

Even as the first Type C locomotives were being built, BR ordered a further 137 examples (D11-147) in January 1959 with production split between Derby Works (D11-67) and Crewe Works (D68-147). This batch of locomotives (subsequently classified Class 45) was fitted with an uprated 12LDA28 engine; designated 12LDA28-B this was rated at 2500 hp @ 750 rpm and was the most powerful single engine then available. In December 1959 a further order was placed for 66 locomotives (D148-199; D1500-13) but specified Brush electrical equipment due to delays with the delivery of Crompton Parkinson equipment hence their later classification as Class 46. Following concerns by BR's North Eastern Region, the order for D194-199; D1500-1513 was cancelled in May 1961 and the equipment allocated to a new Co-Co design (later Class 47) that was under development by J.F. Harrison (BR's Chief Mechanical Engineer) while the final ten Class 45 locomotives (D138-147) were transferred to the Class 46 order. The final Class 45 locomotive was D57 (later 45042) which was initially fitted with an uprated 12LDA28 engine; designated 12LDA28-C, the engine was rated at 2750 hp @ 800 rpm and intended for fitment into the new Class 47 design, hence once the trial was completed the engine was replaced by a standard 12LDA28-B unit.

The introduction of the TOPS renumbering scheme in 1974 allowed the three variants to be clearly identified. D1-10 with the original 12LDA28 engine rated at 2300 hp and nose-mounted discs became Class 44, D11-137 with the uprated 12LDA28-B engine rated at 2500 hp and discs replaced by nose-mounted headcode boxes became Class 45, and D138-193 with the uprated 12LDA28-B engine, nose-mounted headcode boxes and Brush electrical equipment became Class 46.

The building of Class 45 locomotives had initially been shared between Derby Works (D11-67) and Crewe Works (D68-137) but the construction of D50-67 was transferred to Crewe in two tranches once Class 46 construction began at Derby. Further changes during construction saw the nose-mounted corridor connection retained only for D11-15, the pair of nose-mounted 2-character headcode boxes being carried by D16-30; D68-107; the nose-mounted central 2-character headcode boxes being carried by D31-67; D108-173 then D174 onwards carrying a central nose-mounted 4-character headcode box. Repairs to accident damage during the 1960s saw the final 4-character headcode box being fitted to random locomotives, then, in the 1970s, the headcode boxes were replaced by nose-mounted marker lights thus making identification of the individual classes more difficult (see **Section 4**).

Despite their bulk and heavy weight, the three classes proved popular with crews and found usage over a large portion of the BR network but especially on the Midland Main Line (MML) services emanating from London St Pancras. My experience of the three classes began with the sight of D72 (later 45050) working a Sunday Manchester Central–St Pancras service through Corby in December 1960 as class members began their takeover of MML services – both freight and passenger – hence this album is my tribute to the three classes as they transformed the 'Cinderella' MML into a premier service between St Pancras and the East Midlands.

Any errors, however, remain mine hence any corrections to the author should be directed through the publishers in the first instance.

Fred Kerr August 2022.

SECTION 1
CLASS 44: D1 - D10 / 44001 - 44010

1.1: MAIN LINE WORKINGS

When delivered during 1959 this class was initially allocated to West Coast Main Line (WCML) depots as BR sought to implement the Clean Air Act 1956 by using diesel traction on named expresses into/out of London. In late 1959/early 1960 D3 - D10 were based at Derby for tests on the Midland Region's (MR) London St Pancras–Derby–Manchester route where their traction characteristics quickly proved well suited to the MR's hilly routes. The trial locomotives returned to the WCML after a short stay of only three months, but in May 1962, following completion of deliveries of the English Electric Class 40 locomotives allocated to the WCML, the fleet was transferred to the MR based at Toton. By that time the Class 45 deliveries had taken over the MR's passenger services and much of the freight services hence the ten Class 44 locomotives were confined to freight workings which they continued powering on an out-and-home basis from Toton until their withdrawal in the late 1970s.

Left (Image 1): *Class doyen D1* Scafell Pike *powers past Corby Sidings on 21 May 1965 while working a Toton–Wellingborough coal service.*

Below (Image 2): *D6* Whernside *powers past Corby Sidings on 6 June 1965 while working a Toton–Wellingborough coal service.*

(Image 3): *44007 (ex D7)* Ingleborough *approaches Kettering station on 1 August 1975 while working a Wellingborough–Toton coal service.*

(Image 4): *The driver of D8* Penyghent *looks back to the shunter at Corby Lloyds Sidings on 8 April 1969 as he shunts his Toton–Lloyds Sidings train into the sidings laden with raw materials for use in the blast furnaces of Corby Steelworks.*

In October 1977 the Diesel & Electric Group (D&EG) sought to raise funds to support the preservation of Class 35 Hymek D7017 by organising a Class 44 'Farewell' railtour from London St Pancras to Manchester Piccadilly via Kettering, Manton, Loughborough, Erewash Valley and Hope Valley (i.e. routes worked by the locomotives with services from Toton). BR demurred on the basis that enthusiasts would not support a charter train hauled by non-heat locomotives and introduced the DAA (Daft as A*******s) Travel Group who had requested a similar itinerary with the thought that the combined approach might fill a complete train. To BR's surprise the train not only filled quickly but attracted sufficient custom to justify a second train. BR's 'apology' was to guarantee the use of all six remaining class members on the two tours.

The first tour ran on 1 October 1977 using 44005 (ex D5) *Cross Fell* on the St Pancras–Toton stage, 44008 (D8) *Penyghent* on the Toton–Manchester–Toton stage and 44009 (ex D9) *Snowdon* on the final Toton–St Pancras stage.

Opposite above (Image 5): *44005 Cross Fell awaits departure from St Pancras*

Opposite middle (Image 6): *44008 Penyghent makes a photo-stop at Bamford on the return journey.*

Opposite below (Image 7): *44009 Snowdon stands at St Pancras after arriving with the tour and the stock moved from the platform. Note that the original nose-mounted corridor and discs have been replaced by a replacement nose with central 4-character headcode box – presumably during accident repairs.*

Above (Image 8): *44008 Penyghent makes a photo-stop at Penistone on the outward journey.*

The 2nd tour ran on 15 October 1977 when the same itinerary and routing was followed. On this occasion 44002 (ex D2) *Helvellyn* powered the St Pancras–Toton stage, 44004 (ex D4) *Great Gable* powered the Toton–Manchester–Toton stage and 44007 (ex D7) *Ingleborough* powered the final Toton–St Pancras stage.

Of interest during this tour was the performance of 44002 *Helvellyn* which had been the locomotive temporarily uprated to 2500 hp and authorised to operate on the WCML at 100 mph in 1964 for pre-electrification trials and performance monitoring. Its performance on this tour suggested that 44002 *Helvellyn* was still operating at 2500 hp despite it having been officially de-rated to 2300 hp once the trials had been completed.

Opposite above (Image 9):
44004 Great Gable *makes a photo-stop at Penistone on the outward journey.*

Opposite middle (Image 10):
44004 Great Gable *stables in Manchester Piccadilly during a stock move.*

Opposite below (Image 11):
44007 Ingleborough *stands in St Pancras after its arrival with the railtour.*

Right (Image 12): *44002* Helvellyn *stands in St Pancras awaiting departure with the railtour as a Class 45/1 locomotive awaits departure with a service to the East Midlands.*

Below (Image 13): *The view from the cab of 44004* Great Gable *as it approaches New Mills Tunnel.*

Left (Image 14): *44010 (ex D10)* Tryfan *is noted stored on Derby Works during a visit on 18 March 1978 while awaiting a decision re its disposal.*

Below (Image 15): *44009 (ex D9)* Snowdon *is noted stored awaiting disposal on Toton Depot during an Open Day visit on 8 June 1979.*

Opposite above (Image 16): *44008 (ex D8)* Penyghent *gives cab rides around Toton Depot on 8 June 1979 during a depot open day.*

Opposite middle (Image 17): *44008 (ex D8)* Penyghent *stables on Toton Depot during a visit on 1 May 1977.*

Opposite below (Image 18): *44007 (ex D7)* Ingleborough *and 44004 (ex D4)* Great Gable *stable inside Toton Depot during a visit on 1 May 1977.*

1.2: PRESERVATION

Two class members entered preservation – D4 / 44004 *Great Gable* normally resident at the Swanwick base of the Midland Railway Centre and D8 / 44008 *Penyghent* normally resident at the Rowsley base of Peak Rail. Both locomotives have visited other heritage lines as 'guest locomotives' for Gala events hence may be noted at various centres.

Left (Image 19): 44004 Great Gable *bears the guise of class doyen D1* Scafell Pike *as it approaches Ramsbottom (East Lancashire Railway) on 4 July 1998 while working the 10:25 Rawtenstall–Bury Bolton St service.*

Below (Image 20): 44004 Great Gable *bears the guise of class doyen D1* Scafell Pike *as it powers out of Summerseat (East Lancashire Railway) on 7 July 1998 while working the 09:40 Bury Bolton St–Rawtenstall service.*

Right (Image 21): *44004* Great Gable *bears the guise of class doyen D1* Scafell Pike *as it approaches Brooksbottom Tunnel (East Lancashire Railway) on 4 July 1998 while working a Bury Bolton St–Rawtenstall service.*

Below (Image 22): *44008 bears the local name* Schiehallion *when stored at Boat of Garden (Strathspey Railway) on 12 August 1985 while awaiting restoration to working order.*

Left (Image 23): *D4 Great Gable stables at Loughborough Central (Great Central Railway) on 7 May 1995 while awaiting its next duty.*

SECTION 2
CLASS 45: D11 - 137 / 45001 - 45077; 45101 - 45150

Until the introduction of the TOPS (Total Operations Processing System) computer system in the 1970s, the fleet of 193 locomotives were simply identified as 'Peaks' following the naming of the first ten (D1-D10 / 44001-44010) after famous UK mountain peaks. When TOPS fleet numbers were allocated, the production series of 127 locomotives became Class 45. The class had been built by both Derby (D11-D49) and Crewe Workshops (D50-D137); D50-D67 had initially been allocated for construction by Derby Works but were later transferred to Crewe Works in two tranches, resulting in minor variations between the two sites. Derby-built D11-D15 retained nose-mounted corridor connections hence the pair of 2-character headcode boxes were mounted either side of this but when the corridor connection was removed the location of the headcode boxes remained for D16-D30 while Crewe built D68-D107 without corridor connections and the headcode boxes also mounted on the nose sides. In an initial design change D31-67 and D108-D137 had them relocated to the centre of the nose as 2 x 2-character headcode boxes. In a later design change locomotives from D174 (Class 46 – see **Section 3**) onwards were built with centrally positioned nose-mounted 4-character headcode boxes. During the 1960s when locomotives received either accident repairs or works attention this final design of headcode box was applied as part of the attention/accident repair works.

When introduced to service the Derby-built locomotives were initially allocated to Derby but then quickly re-allocated to ex-Midland Railway depots. D11-31 were initially transferred to the diesel depot at (Leeds) Neville Hill then shortly after to (Leeds) Holbeck to power ex-Midland Railway services north to Carlisle and south to London St Pancras; D32-D42 were transferred to Bristol (Bath Road) to work cross-country services on the North East–South West route between Bristol and Newcastle while D43-56; D58-D137 remained at Derby for working both freight and passenger services on the Midland Main Line (MML). D57 (later 45042) was experimentally fitted with a 12LDA28-C engine uprated to 2750 hp @ 800 rpm for a short period when initially built. The experiment lasted around two years before the 12LDA28-C engine was replaced by a standard 12LDA28-B engine and transferred to one of the Class 47 locomotives then under construction.

These allocations were changed in September 1963 when BR adopted allocation to areas and the MML Class 45 fleet was allocated to one of four areas – D14 (based on Cricklewood), D15 (based on Leicester), D16 (based on Toton) and Midland Lines (ML) as common user by any of the new areas.

A further change arose in April 1966 when the opening of the WCML electrification to Euston saw the London Midland Region exchange its London–Leeds/Bradford services for the Eastern Region's London–Sheffield services while retaining *The Thames-Clyde Express* (St Pancras–Glasgow Central) and *The Waverley* (St Pancras–Edinburgh Waverley) daytime and overnight sleeper services.

2.1: PRE-TOPS = D11 - D137

(Image 24): *D12 powers past Corby Sidings on 20 May 1965 while working a Nottingham–St Pancras service. This was one of the 5 Derby-built locomotives (D11-15) that were fitted with nose-mounted corridor connections and 2 x 2-character headcode boxes while D16 onwards were built without corridor connections; the corridor connections were subsequently removed during overhauls/accident repair works.*

(Image 25): *D118 approaches Corby North on 2 April 1968 while working a Wellingborough–Cargo Fleet ironstone service.*

(Image 26): *The driver of D126 looks back to watch the shunter on 11 February 1967 as he prepares to shunt his Toton–Corby Lloyds Sidings coal service into Lloyds Sidings for use in Corby Steelworks blast furnaces.*

(Image 27): *D136 powers past Corby Sidings on 8 July 1971 while working a Wellingborough–Toton mixed goods service. Note that the nose-mounted 2 x 2-character headcode boxes have been replaced by a single nose-mounted 4-character headcode box.*

(Image 28): *D70 The Royal Marines propels a guards van as it passes Kettering on 6 April 1968 while en route from Wellingborough to Corby Sidings to start local ironstone trip workings.*

(Image 29): *D15 powers through Newton Abbot on 16 October 1970 while working a Sheffield–Plymouth cross-country service. Note that its original pair of split 2-character headcode boxes and nose-mounted corridor gangway connection have been replaced by a nose-mounted 4-character headcode box.*

(Image 30): *D76 powers past Corby Sidings on 22 June 1965 while working a Nottingham–St Pancras service. Note that this Crewe-built locomotive had been built without corridor gangway connections yet had the pair of nose-mounted 2-character headcode boxes.*

(Image 31): D82 powers past Corby North signalbox on 3 April 1968 while working a Wellingborough–Hartlepool ironstone service. Note that its original nose-mounted 2-character headcode boxes have been replaced by a central nose-mounted 4-character headcode box.

(Image 32): D81 approaches Kettering on 6 April 1968 while working a Nottingham–St Pancras service.

Opposite above (Image 33): *D105 drifts past Corby Lloyds South on 8 April 1969 while working a Wellingborough–Cargo Fleet ironstone service. Note that the locomotive is painted in the new corporate blue livery which had been applied by hand at Toton Depot as the first class member to receive the new corporate livery.*

Opposite middle (Image 34): *D98* Royal Engineer *powers past Corby North signalbox on 3 April 1968 while working a Colwick–Wellingborough coal service.*

Opposite below (Image 35): *D56* Bedfordshire & Hertfordshire Regiment (TA) *passes Corby Sidings on 10 June 1965 while working an Earles Sidings–Wellingborough cement service.*

Above (Image 36): *D79 eases out of Corby Lloyds Sidings on 8 July 1971 having deposited a Toton–Corby Lloyds Sidings portion and now continuing southwards with a Toton–Northfleet coal train for use in the cement works there. Note the second man acknowledging the guard's signal that the complete train had started moving.*

2.2: TOPS

Class 45s received their TOPS numbers from 1974 but the number sequence was complicated by the decision to convert fifty locomotives to Class 45/1 by the fitting of Electric Train Heating (ETH) when they received works attention. This resulted in TOPS identifying two classes – Class 45/0 for locomotives with/without steam heating nominally restricted to freight duties and Class 45/1 with ETH nominally restricted to passenger services out of St Pancras. The 45/1 sub-class was later re-allocated between (Leeds) Holbeck and (Sheffield) Tinsley depot when replaced by High Speed Trainsets (HST) for use on Trans-Pennine passenger services including Newcastle–Liverpool and Scarborough–Liverpool/North Wales. Once the Class 45/1 locomotives had been renumbered and the first of the Class 45/0 renumbering had taken place, the remaining Class 45/0 locomotives were renumbered in sequence of the outstanding original numbered locomotives beginning with D17/45024.

2.2.1: CLASS 45/0 45001 - 45077
2.2.1.1: Freight Services

(Image 37): 45071 (ex D125) powers past Corby Lloyds North signalbox on 29 March 1978 while working a Toton–Northfleet MGR service of coal for the Kent cement works. Note the nose-mounted marker lights which were first applied to the locomotive when released from accident repair by Derby Works during late 1975; a modified design of lights was subsequently applied to both Class 45 and 46 locomotives over the next couple of years.

(Image 38): *45014 (ex D137) The Cheshire Regiment eases its Cricklewood–Cliffe Hill civil engineering train across the lines at Glendon North Junction on 4 January 1984. This point, at 75 miles north of London St Pancras, marks the end of the longest 4-track route out of any London terminus and the train is crossing from the 4-track to the 2-track section to head north towards Leicester.*

(Image 39): *45070 (ex D122) eases through the derelict Corby station platform on 31 July 1981 while working a Toton–Wellingborough mixed goods service.*

(Image 40): *45033 (ex D39) pilots Class 45/1 45146 (ex D66) as they cross onto the 2-track section at Glendon North Junction on 23 March 1982 while working a Ridiham Dock–Tyne Yard MGR service.*

(Image 41): *45049 (ex D71) The Staffordshire Regiment (The Prince of Wales's) powers past Corby Lloyds North signalbox on 29 March 1978 while working a Toton–Acton coal service.*

(Image 42): *45065 (ex D110) drifts past Bedford Kempston Road on 23 February 1979 while working a Toton–Acton mixed goods service.*

(Image 43): *45060 (ex D100) Sherwood Forester climbs out of the Welland Valley past Gretton on 19 September 1984 while working a Toton–Corby BSC trip service.*

(Image 44): *45034 (ex D42) pilots Class 47/0 47258 (ex D1938) as they climb through Peak Forest on 25 April 1987 while working a Tunstead–Margam limestone service.*

(Image 45): *45010 (ex D112) shunts the Civil Engineer's Tunnel Inspection Trainset into Corby station sidings on 7 October 1984 after completing an inspection of Corby Tunnel.*

(Image 46): *45004 (ex D77)* Royal Irish Fusilier *powers past Corby North signalbox on 4 June 1982 while working a Toton–Acton service.*

(Image 47): *45059 (ex D98)* Royal Engineers *passes Corby Lloyds North signalbox on 29 March 1978 while working an Acton–Toton mixed goods service.*

2.2.1.2: Passenger Services

After displacement from MML services, class members were re-allocated between (Leeds) Holbeck and (Sheffield) Tinsley to work a variety of services in the north of England particularly on trans-Pennine services.

Opposite above left (Image 48): *45009 (ex D37) was withdrawn from service but a shortage of Type 4 locomotives saw it being one of the class members that was overhauled and restored to service hence its appearance at (Liverpool) Fazakerley PW yard on 15 May 1983 while working a track train from the nearby Orrel Park engineering site.*

Opposite above right (Image 49): *45013 (ex D20) stands in Southport station on 11 March 1982 while awaiting departure with the Dean Lane–Appley Bridge GMC Wasteliner service after its reversal in Bradford Sidings.*

Opposite below (Image 50): *45005 (ex D79) stands in Manchester Victoria in the early hours of 15 October 1983 while awaiting departure with the 00:35 mail/newspaper service to Newcastle.*

Above (Image 51): *Although nominally restricted to freight duty, Class 45/0 locomotives continued to see use on passenger duty as on 22 March 1981 when 45049 (ex D71) The Staffordshire Regiment (The Prince of Wales's) powered out of St Pancras while working a service to Derby.*

(Image 52): *45035 (ex D44) curves past Desborough on 12 May 1979 while working a Derby–St Pancras service.*

(Image 53): *45043 (ex D58)* The King's Own Royal Border Regiment *awaits departure from Kettering on 2 April 1979 while working a Derby–St Pancras service.*

Above (Image 54): *45044 (ex D63) Royal Inniskilling Fusilier speeds through Rushton on 9 August 1975 while working a St Pancras–Nottingham service.*

Right (Image 55): *45059 (ex D98) Royal Engineers awaits departure from St Pancras on 20 March 1976 with a Football charter (Footex) to Nottingham.*

Below (Image 56): *45054 (ex D95) pilots failed Class 45/1 45116 (ex D47) past Glendon North Junction on 27 October 1979 while working a St Pancras–Nottingham service.*

Once displaced from MML services out of St Pancras, class members were transferred to Sheffield (Tinsley) and (Leeds) Holbeck depots to work services in the north west of England including trans-Pennine services to Newcastle and Scarborough.

(Image 57): *45026 (ex D21) curves through Birkett Common on 24 April 1984 while working a Leeds–Carlisle service.*

(Image 58): *45046 (ex D68)* The Royal Fusilier *enters Manchester Victoria on 19 June 1982 while working a Scarborough–Liverpool Lime St service.*

(Image 59): *45007 (ex D119) approaches Liverpool Lime St on 25 April 1982 while working a service from Newcastle.*

(Image 60): *45026 (ex D21) weaves out of York on 11 August 1983 while working a Scarborough–Liverpool Lime St service.*

2.2.1.3: Preservation

After withdrawal from service, three class members entered preservation – 45015 (ex D14) stored on the Battlefield Line where an ownership dispute has seen it slowly rot away; 45041 (ex D53) *Royal Tank Regiment* normally based on the Battlefield Line and 45060 (ex D100) *Sherwood Forester* normally based at Barrow Hill. The latter pair also visit other heritage lines as 'guest locomotives' on occasion for gala events.

Left (Image 61): *45041* Royal Tank Regiment *stables at the Swanwick base of the Midland Railway Centre on 5 August 1997.*

Below (Image 62): *45041* Royal Tank Regiment *stands as a display item at Crewe Works Open Day on 10 September 2005.*

(Image 63): *45041* Royal Tank Regiment *weaves out of Kidderminster (Severn Valley Railway) on 18 May 2018 while working the 09:55 Kidderminster–Bridgnorth service.*

(Image 64): *45041* Royal Tank Regiment *ascends Eardington Bank (Severn Valley Railway) on 17 May 2018 while working the 12:59 Kidderminster–Bridgnorth service.*

Above (Image 65) / Below (Image 66): *45060 (ex D100)* Sherwood Forester *pilots Class 45 45135 (ex D99)* 3rd Carabinier *as they call at Irwell Vale (East Lancashire Railway) on 6 July 2002 while working the 10:45 Bury-Bolton St–Irwell Vale service in top 'n tail mode with Class 33/2 33201.*

Right (Image 67): *The* Sherwood Forester *nameplate carried by 45060 (ex D100).*

Below (Image 68): *45060 (ex D100)* Sherwood Forester *displays its restored pair of 2-character headcode boxes as it approaches Ramsbottom (East Lancashire Railway) on 9 July 1998 while working the 09:35 Rawtenstall–Bury Bolton St service and passes Class 46 D172* Ixion *stabled as standby locomotive.*

(Image 69): *45060 (ex D100) Sherwood Forester poses in the Barrow Hill display yard on 6 October 2002 during a gala event.*

(Image 70): *45060 (ex D100) Sherwood Forester calls at Summerseat (East Lancashire Railway) on 2 July 2010 while working the 13:36 Rawtenstall–Heywood service.*

(Image 71): *45060 (ex D100) Sherwood Forester stables in the Harry Needles storage area at Barrow Hill on 5 October 2003 during a steam locomotive gala event.*

(Image 72): *45060 (ex D100) Sherwood Forester displays its restored pair of 2-character headcode boxes (compare with **Image 69**) and* THE ROBIN HOOD *headboard, which it carried when regularly powering that weekday St Pancras–Nottingham service in the early 1960s, while on display at Barrow Hill on 8 August 2009.*

2.2.2: CLASS 45/1 45101 - 45150
2.2.2.1: Passenger Service
Scenes at London St Pancras

Above (Image 73): *45112 (ex D61) The Royal Army Ordnance Corps arrives on 20 March 1976 while working a service from Nottingham.*

Opposite above (Image 74): *Peak hour departures on 18 February 1977 sees (left) 45116 (ex D47) awaiting departure with the 17:23 MASTER CUTLER service to Sheffield as 45149 (ex D135) waits to follow with the 17:26 service to Derby.*

Opposite below (Image 75): *45138 (ex D92) awaits departure on 20 April 1976 with the 17:26 service to Derby.*

(Image 76): *45126 (ex D32) eases through the station throat on 24 June 1983 while working a stock move to Cricklewood Carriage Sidings.*

(Image 77): *45131 (ex D124) powers out on 22 March 1981 while working a service to Sheffield.*

(Image 78): *45135 (ex D99)* 3rd Carabinier *powers out on 22 March 1981 while working a service to Sheffield.*

Opposite above (Image 79): 45150 (ex D78) departs on 22 March 1981 while working a service to Sheffield.

Opposite below (Image 80): 45103 (ex D116) departs on 24 June 1983 while working a stock move to Cricklewood Carriage Sidings.

Right (Image 81): 45106 (ex D106) approaches on 12 January 1982 while working the 05:55 Derby–St Pancras service.

Below (Image 82): 45114 (ex D94) stables in Cambridge St Refuelling Point on 23 March 1977 while awaiting its next duty.

Scenes around Luton

(Image 83): *45113 (ex D80) calls on 2 June 1979 while working a Derby–St Pancras service.*

(Image 84): *45104 (ex D59)* The Royal Warwickshire Fusiliers *passes on 3 June 1977 while working a Sheffield–St Pancras service.*

Scenes around Bedford

Left (Image 87): *45125 (ex D123) stables in Bedford Carriage Sidings on 13 March 1983 during a period when a lack of Class 127 BedPan trainsets saw some Bedford–St Pancras services operated with locomotive + coaches trainsets.*

Below (Image 88): *45139 (ex D109) departs from Bedford Midland on 13 August 1975 while working a Manchester Piccadilly–St Pancras service.*

(Image 85): 45107 (ex D43) approaches at speed on 3 June 1977 while working THE MASTER CUTLER (17:23 St Pancras–Sheffield) service.

(Image 86): 45109 (ex D85) departs on 3 June 1977 while working a Derby–St Pancras service.

Scenes at Kettering South Junction

(Image 89): *45115 (ex D81) races through the rain on 5 June 1982 while working a St Pancras–Sheffield service.*

(Image 90): *45107 (ex D43) speeds past on 5 June 1982 while working a St Pancras–Sheffield service.*

Scenes around Kettering

(Image 91): *45102 (ex D51) departs on 2 April 1979 while working a Nottingham–St Pancras service.*

(Image 92): *45105 (ex D86) calls on 27 September 1980 while working a St Pancras–Derby service.*

(Image 93): *45141 (ex D82) departs on 10 January 1982 while working a Nottingham–St Pancras service.*

(Image 94): *45119 (ex D34) calls at Kettering on 26 January 1979 while working a St Pancras–Derby service. Note this locomotive was officially the 1000th main line diesel locomotive delivered to BR under the Modernisation Plan.*

(Image 95): *45104 (ex D59)* The Royal Warwickshire Fusiliers *curves past Kettering Iron & Coal Sidings on 10 March 1983 as it heads north out of Kettering towards Glendon while working a St Pancras–Nottingham service.*

Scenes around Glendon Junction

(Image 96): *45140 (ex D102) passes on 11 March 1980 while working a Nottingham–St Pancras service.*

(Image 97): *45149 (ex D135) passes on 11 March 1980 while working a Sheffield–St Pancras service.*

Scenes around Glendon Junction

Opposite above (Image 98): *45129 (ex D111) approaches on 11 March 1980 while working a St Pancras–Nottingham service.*

Opposite below (Image 99): *45110 (ex D73) approaches on 2 April 1977 while working a St Pancras–Leeds service.*

Above (Image 100): *45124 (ex D28) approaches on 4 August 1979 while working a Nottingham–Ramsgate summer dated service.*

Scenes around Glendon North Junction

This location, at 75 miles north of St Pancras, marked the end of the longest 4-track main line from a London terminus.

(Image 101): *45110 (ex D73) curves past on 27 October 1979 while working a Nottingham–St Pancras service.*

(Image 102): *45118 (ex D67)* The Royal Artilleryman *curves past on 23 March 1982 while working a Sheffield–St Pancras service.*

(Image 103): *45120 (ex D107) curves past on 23 March 1982 while working a St Pancras–Nottingham service.*

(Image 104): *45139 (ex D109) passes on 23 March 1982 while working a St Pancras–Nottingham service.*

(Image 105): *45114 (ex D94) crosses from the 4-track section onto the 2-track section on 27 March 1976 while working a St Pancras–Liverpool Lime St service diverted from Euston due to WCML engineering works. This service was the first out of St Pancras to introduce Mk III coaches to the Midland Main Line.*

Scenes around Glendon North Junction

(Image 106): *45134 (ex D126) approaches on 23 March 1982 while working a Sheffield–St Pancras service.*

(Image 107): *45143 (ex D62)* 5th Royal Inniskilling Dragoon Guards *curves past on 27 March 1976 while working a St Pancras–Derby service.*

Scenes around Rushton

Right (Image 108): *45146 (ex D66) passes on 9 August 1975 while working a St Pancras–Derby service.*

Below (Image 109): *45106 (ex D106) approaches on 10 January 1982 while working a St Pancras–Nottingham service.*

Scenes around Rushton

Above left (Image 110): *45150 (ex D78) approaches on 9 August 1975 while working a Sheffield–St Pancras service.*

Above right (Image 111): *45117 (ex D35) approaches on 9 August 1975 while working a Sheffield–St Pancras service.*

Below (Image 112): *45117 (ex D35) passes on 10 January 1982 while working a Sheffield–St Pancras service.*

Scenes around Desborough

Right (Image 113): *45143 (ex D62)* 5th Royal Inniskilling Dragoon Guards *passes on 5 May 1979 while working a St Pancras–Manchester Piccadilly service.*

Below (Image 114): *45105 (ex D86) breasts Desborough Bank on 12 May 1979 while working a Sheffield–St Pancras service.*

Bottom (Image 115): *45144 (ex D55)* Royal Signals *passes on 5 May 1979 while working a St Pancras–Sheffield service.*

Scenes around Desborough

Left (Image 116): *45145 (ex D128) passes on 28 March 1981 while working a Nottingham–St Pancras service.*

Below (Image 117): *45115 (ex D81) passes on 28 March 1981 while working a Nottingham–St Pancras service.*

Bottom (Image 118): *45147 (ex D41) passes on 12 May 1979 while working a Sheffield–St Pancras service.*

Right (Image 119): *45139 (ex D109) breasts Desborough Bank on 5 May 1979 while working a Derby–St Pancras service.*

Below (Image 120): *45129 (ex D111) breasts Desborough Bank on 5 May 1979 while working a Derby–St Pancras service.*

Bottom (Image 121): *45137 (ex D56) The Bedfordshire & Hertfordshire Regiment (TA) passes on 12 May 1979 while working a Derby–Cricklewood mail/newspaper empty stock service.*

2.2.2.1.1: Diversion via Manton Line

The signalbox at Glendon Junction controlled access to the 'Manton Line' which is the regular diversionary route when engineering work takes place on the MML between Kettering and Leicester. After diverting at Glendon Junction the route continues to Manton where it joins the Syston–Peterborough Railway from Peterborough to Leicester. The 'Manton Line' had been the original St Pancras–Nottingham main line and it diverged at Melton Mowbray to continue to Nottingham via Old Dalby; in the Beeching-led restructuring of the 1960s the line was closed between Melton Mowbray and Nottingham and services operated to Syston Junction thence north to Loughborough or south to Leicester. The closed route to Nottingham was subsequently resuscitated in part to create the Asfordby test track for Derby's Railway Technical Centre (RTC) while the Glendon Junction–Melton Mowbray–Syston Junction line reverted to its original role as a freight line but maintained for diversions from the MML route between Kettering and Leicester.

Above (Image 122): 45108 (ex D120) approaches the derelict Corby station on 6 June 1982 while working a diverted St Pancras–Derby service.

Opposite above (Image 123): 45132 (ex D22) passes Corby Lloyds Sidings on 20 May 1979 while working a diverted Nottingham–St Pancras service.

Opposite below (Image 124): 45120 (ex D107) passes Corby Steelworks, adjacent to Lloyds Sidings, on 29 March 1978 while working a St Pancras–Sheffield service, diverted due to a broken rail on the main line between Kettering and Leicester.

Corby steelworks was a major source of traffic for this line and required two flights of sidings – Lloyds Sidings to receive the raw materials and Corby Sidings to despatch the finished product of steel tubes. The traffic has since declined with closure of the steelworks in the 1970s, hence the major traffic in 2021 is the daily train of steel strip from Margam to make the steel tubes which was the original finished product of the steelworks.

(Image 125): *45125 (ex D123) approaches Corby Lloyds South on 6 March 1977 while working a diverted St Pancras–Nottingham service.*

(Image 126): *45130 (ex D117) passes Corby North signalbox on 2 September 1984 while working a diverted Derby–St Pancras service.*

A major structure on the diversion route is Harringworth Viaduct which carries the railway across the Welland Valley hence also being called Welland Viaduct or Seaton Viaduct. The viaduct is 1275 yards long and comprises 82 arches, each with a 40 foot span, hence is the longest masonry viaduct across a valley in the United Kingdom.

(Image 127): *45136 (ex D88) crosses the viaduct on 16 September 1984 while working a diverted Derby–St Pancras service.*

(Image 128): *45117 (ex D35) crosses the viaduct on 3 May 1981 while working a diverted Nottingham–St Pancras service.*

(Image 129): *45148 (ex D130) stands in Euston on 20 October 1979 with a Sheffield–St Pancras service diverted from Bedford via Bletchley due to the commissioning of West Hampstead Power Box.*

2.2.2.1.2: Northern Climes

Once displaced from MML services, class members were transferred to the north of England to work services in the north and north west including Trans-Pennine passenger services to Newcastle and Scarborough.

Left (Image 130): *45117 (ex D35) and 45113 (ex D80) stable outside (Sheffield) Tinsley depot on 15 July 1982.*

Opposite above (Image 131): *45103 (ex D116) curves through the battlements of Conway Castle on 4 April 1986 while working a Scarborough–Bangor service.*

Opposite below (Image 132): *45107 (ex D43) curves onto Miles Platting Bank on 19 June 1982 as it departs from Manchester Victoria while working a Liverpool Lime St–Newcastle service.*

(Image 133): *45109 (ex D85) awaits departure from Liverpool Lime St on 4 April 1986 while working a Liverpool–Scarborough service.*

(Image 134): *45115 (ex D81) awaits departure from Manchester Victoria on 13 May 1985 while working the 02:15 newspaper/mail service to Leeds.*

(Image 135): *Peaks meet at Liverpool Lime St on 30 November 1984 as 45060 (ex D100)* Sherwood Forester *makes a station shunt move as 45101 (ex D96) awaits departure with a service to Newcastle.*

(Image 136): *45122 (ex D11) stables in Liverpool Lime St station on 13 November 1982 while awaiting its next Trans-Pennine duty.*

(Image 137): *45128 (ex D113) stables with a sister locomotive at (Sheffield) Tinsley depot on 15 July 1982.*

(Image 138): *45144 (ex D55) Royal Signals reverses the stock of a Manchester Victoria–Southport service into Southport on 7 July 1987 as a prelude to working a Southport–Manchester Victoria service in the early days of covering the shortage of DMU trainsets by operating locomotive-hauled trainsets.*

(Image 139): *45123 (ex D52) The Lancashire Fusilier calls at Manchester Victoria on 22 August 1984 while working a Liverpool Lime St–Scarborough service.*

(Image 140): *45133 (ex D40) powers past Lowton Junction on 18 June 1983 while working a Liverpool Lime St–Scarborough service.*

(Image 141): *45127 (ex D87) powers past Lowton Junction on 18 June 1983 while working a Liverpool Lime St–Scarborough service.*

(Image 142): *45101 (ex D96) approaches York on 11 August 1983 while working a Liverpool Lime St–Scarborough service.*

(Image 143): *45106 (ex D106), bearing green unlined livery and depot embellishments to mark it being one of the last class members to remain in active service, stands in Southport on 9 September 1988 after arriving from (Wigan) Springs Branch depot as part of a driver familiarisation run in preparation for working stock shunt duties at Southport in connection with steam locomotive charters that operated a few days later. On the day it became unavailable after derailing on Springs Branch depot and was replaced by Class 20 20019 + 20065.*

Scenes from the Settle & Carlisle route

(Image 144): *45109 (ex D85) curves through Helm on 31 March 1984 while working a charter from Cleethorpes to Carlisle.*

(Image 145): *45122 (ex D11) curves through Armathwaite on 29 December 1983 while working a Leeds–Carlisle service.*

(Image 146): *45142 (ex D83) curves through Dent on 20 April 1984 while working a Carlisle–Leeds service.*

Right (Image 147): *45142 (ex D83) breasts Ais Gill Summit on 21 April 1984 while working a Carlisle–Leeds service.*

Below (Image 148): *45150 (ex D78) powers past Ais Gill Summit on 21 April 1984 while working a Leeds–Carlisle service.*

(Image 149): *45128 (ex D113) takes a break from passenger duty on 5 April 1984 while working an Aintree FLT–Garston FLT transfer service as it approaches Bootle New Strand.*

(Image 150): *45120 (ex D107) pilots Class 86/2 86235 Novelty over Lowton Junction as they curve onto the WCML at Golborne Junction while working the 05:13 Euston–Stranraer service which had been diverted from Crewe via Manchester and Chat Moss due to engineering works between Crewe and Golborne.*

2.2.2.2: Preservation

Nine class members entered preservation after withdrawal from active service – 45105 (ex D86) normally based at Barrow Hill, 45108 (ex D120) normally based at the Nottingham Transport Museum, 45112 (ex D61) *The Royal Army Ordnance Corps* normally based at the Nemesis Rail site at Burton on Trent, 45118 (ex D67) *The Royal Artilleryman* normally based at Barrow Hill, 45125 (ex D123) normally based at the Loughborough site of the Great Central Railway, 45132 (ex D22) normally based at the Epping & Ongar Railway, 45133 (ex D40) normally based at the Swanwick site of the Midland Railway Centre, 45135 (ex D99) *3rd Carabinier* normally based at the East Lancashire Railway, and 45149 (ex D135) normally based at the Toddington site of the Gloucestershire Warwickshire Railway.

Right (Image 151): *45105 (ex D86) stables in Barrow Hill Roundhouse on 23 August 2008 while under restoration to working order replete with nose-mounted 4-character headcode box.*

Below (Image 152): *45108 (ex D120) stables at the Swanwick site of the Midland Railway Centre on 8 November 2014.*

Scenes of 45108 (ex D120) at work on the East Lancashire Railway include it …

(Image 153): … powering across the River Irwell at Summerseat (East Lancashire Railway) on 24 September 2016 while working the 10:35 Heywood–Rawtenstall service.

(Image 154): … calling at Irwell Vale on 18 February 2017 while working the 15:55 Heywood–Rawtenstall service.

(Image 155): ... forming the rear of the 09:20 Bury Bolton St–Rawtenstall service as it departs from Ramsbottom on 18 February 2017 behind Class 14 D9531 Ernest.

(Image 156): ... approaching Summerseat on 5 July 2019 while working the 11:32 Ramsbottom–Bury Bolton St shuttle service.

(Image 157): ... forming the rear of the 13:36 Rawtenstall–Heywood service on 17 February 2018 as it passes Townsend Fold powered by Class 33/1 33109 Captain Bill Smith RN.

(Image 158): *45108 (ex D120) weaves out of Ramsbottom Siding (East Lancashire Railway) on 7 July 2018 while working a stock move to the station to form a Ramsbottom–Bury Bolton St shuttle service.*

(Image 159): *45112 (ex D61)* The Royal Army Ordnance Corps *is the only preserved Class 45/1 to have received main line certification hence its use on 20 March 2002 to transfer stock from Carnforth (West Coast Railway Company) to Hathersage for an overnight charter from Hathersage to Crewe involving both the Class 45/1 and Coronation Princess Class 4-6-2 6233* Duchess of Sutherland; *the transfer is seen approaching Leyland on the outward journey.*

(Image 160): *45108 (ex D120) approaches Ramsbottom (East Lancashire Railway) on 7 July 2018 while working a stock move from the siding to form a Ramsbottom–Bury Bolton St shuttle service.*

Once its main-line certification lapsed, 45112 (ex D61) The Royal Army Ordnance Corps *was restricted to being either a 'guest' locomotive at heritage line gala events or a display item at railway events, hence its appearance as a display item at …*

Opposite above (Image 161): … Crewe Works Open Day on 10 September 2005.

Opposite below (Image 162): … Barrow Hill on 22 August 2008.

Right (Image 163): 45125 (ex D123) stables at the Loughborough Central site of the Great Central Railway on 15 August 2005 where it has been restored to original condition replete with nose-mounted central 2 x 2-character headcode boxes and bearing the name Leicestershire and Derbyshire Regiment Yeomanry *that was originally carried by Class 46 D163 / 46026.*

(Image 164): *45125 (ex D123) Leicestershire and Derbyshire Yeomanry stables at the Loughborough Central site of the Great Central Railway on 13 November 2008.*

(Image 165): *45133 (ex D40) breasts Eardington Bank (Severn Valley Railway) on 5 October 2012 while working the 09:53 Kidderminster–Bridgnorth service.*

(Image 166): *45133 (ex D40) departs from Rawtenstall (East Lancashire Railway) on 7 July 1998 while working the 13:50 Rawtenstall–Bury Bolton St service.*

Scenes of 45135 (ex D99) 3rd Carabinier *at work on the East Lancashire Railway include it ...*

Left (Image 167):
... approaching Ramsbottom on 5 July 2002 while working the 10:10 Irwell Vale–Bury Bolton St service in top 'n tail mode with Class 45/0 45060 (ex D100) *Sherwood Forester*.

Below (Image 168):
... crossing the River Irwell as it approaches Summerseat on 9 July 2005 while working the 10:25 Rawtenstall–Bury Bolton St service in top 'n tail mode with Class 35 D7076.

(image 169): *... departing from Rawtenstall on 16 March 2002 while working the 17:00 Rawtenstall–Bury Bolton St service.*

(Image 170): *... curving across Burrs Common on 5 July 2003 while working the 16:54 Bury Bolton St–Ramsbottom shuttle service.*

45149 (ex D135) has been restored to working order at the Toddington base of the Gloucestershire Warwickshire Railway where it was photographed awaiting its next duties on …

Left (Image 171):
… 19 August 2013.

Below (Image 172):
… 29 August 2016.

SECTION 3
CLASS 46: D138 - D193 / 46001 - 46056

In the early 1960s the supply of electrical equipment from Crompton Parkinson became subject to delays and the British Transport Commission (BTC) elected to place further orders for the BR Type 4 but specifying Brush electrical equipment; when TOPS renumbering began, this series was classified as Class 46. In essence two batches were supplied – D138-165 / 46001-46028 to the Midland Lines for MML services and D166-193 / 46029-46056 to North Eastern Region depots at (Leeds) Holbeck and Gateshead for Trans-Pennine services from both Newcastle and Scarborough while the Gateshead allocation was also used on secondary ECML services. The headcode boxes were also subject to change with D138-173 having the two nose-mounted central 2-character headcode boxes and D174 onwards having the nose-mounted 4-character headcode box. This final 4-character design was subsequently fitted to earlier examples during accident repair or works overhauls.

With the withdrawal of the Western Region hydraulic classes the consequent transfer of replacement locomotives included the MML batch which were distributed between Bristol Bath Road, Cardiff Canton and Plymouth Laira during the late 1960s and early 1970s to power both freight and passenger services, while the Class 45 locomotives originally based on the Western Region were re-allocated between MML depots and (Leeds) Holbeck.

3.1: PRE - TOPS = D138 - D193

Above (Image 173): D155 approaches Kettering on 6 April 1968 while working a Sheffield–St Pancras service.

Right (Image 174): D163 Leicestershire and Derbyshire Yeomanry sets back a Toton–Corby Lloyds Sidings coal train into the sidings for use in the Corby Steelworks on 11 February 1967.

3.2: TOPS = 46001 - 46056

Opposite above (Image 175): *46013 (ex D150) calls at Exeter St Davids on 3 April 1976 while working a Bristol Temple Meads–Plymouth mail service.*

Opposite below (Image 176): *46026 (ex D163)* Leicestershire and Derbyshire Yeomanry *speeds through Slough on 1 March 1979 while making an Acton Yard–Westbury locomotive move.*

Above (Image 177): *46032 (ex D169) passes Belle Isle on 22 March 1980 while working a Kings Cross–Bounds Green stock move.*

(Image 178): 46050 (ex D187) races past Essendine on 23 August 1974 while working a Scarborough–Kings Cross service.

(Image 179): 46004 (ex D141) speeds out of York past Chaloners Whin on 30 April 1977 while working a Newcastle–Cardiff cross-country service.

(Image 180): *46044 (ex D181) speeds through Fletton as it approaches Peterborough North on 12 April 1980 while working a Kings Cross–Cleethorpes service.*

(Image 181): *46009 (ex D146) races out of Stoke Tunnel on 2 May 1981 while working a Kings Cross–Cleethorpes service.*

(Image 182): An unidentified Class 46 locomotive approaches Glendon South signalbox on 23 March 1982 while working a Mountsorrel–Radlett aggregate service – as viewed from Glendon North Junction signalbox.

(Image 183): 46028 (ex D165) climbs under the flyover as it approaches Corby Tubeworks Siding on 22 February 1982 while working a Lackenby–Corby BSC 'Tubeliner' service.

(Image 184): 46037 (ex D174) powers through the derelict Corby station on 24 January 1984 while working a Toton–Acton coal service.

(Image 185): 46050 (ex D187) curves past Corby Tarmac Siding on 23 March 1982 while working a Corby BSC–Lackenby 'Tubeliner' service.

(Image 186): *46011 (ex D148) stables on Haymarket Depot on 6 April 1980 while awaiting its next duty.*

(Image 187): *46045 (ex D182) awaits departure from Edinburgh Waverley on 5 April 1980 while working an Edinburgh–Kings Cross sleeper service.*

(Image 188): *46029 (ex D166) curves through Selby on 3 September 1981 while working a (Peterborough) New England–Newcastle van service.*

Above (Image 189): *46029 (ex D166) departs from Liverpool Lime St on 17 April 1979 while working a service to Newcastle.*

Opposite above (Image 190): *46051 (ex D188) awaits departure from Newcastle on 9 May 1981 while working a (Peterborough) New England–Edinburgh mail service.*

Opposite below (Image 191): *46026 (ex D163)* Leicestershire and Derbyshire Yeomanry *stables in York Holgate Siding on 11 August 1983 while awaiting its next duty.*

(Image 192): *46011 (ex D148) enters Kings Cross en route from Finsbury Park depot on 10 July 1981 to take up duty on a northbound service.*

(Image 193): *46023 (ex D160) enters Selby on 3 September 1981 while working a Kings Cross–Newcastle service.*

(Image 194): *46030 (ex D167) and 46049 (ex D186) stable overnight at a Kings Cross suburban platform in the early hours of 20 September 1980 while awaiting their next duties.*

(Image 195): *46056 (ex D193) stables in Kings Cross in the early hours of 20 September 1980 after arriving with an Aberdeen–Kings Cross sleeper service.*

(image 196): *46001 (ex D138) stables in Finsbury Park depot on 16 December 1981 shortly after its final withdrawal from service having previously been withdrawn from traffic in Period 12 1980 but restored at Gateshead in Period 9 1981 due to a shortage of Type 4 locomotives.*

3.3: PRESERVATION

Three class members entered preservation – 46010 (ex D147) at Llangollen Railway but subsequently sold to the Nottingham Transport Museum based at Ruddington, 46035 (ex D172) based at the Rowsley base of Peak Rail as part of the Pete Waterman collection, and 46045 (ex D182) normally based at the Swanwick site of the Midland Railway Centre. All three often visit other heritage lines as 'guest locomotive' at gala events or attend railway events as display items.

Right (Image 197):
46010 (ex D147) descends from Dee Bridge (Llangollen Railway) and passes Pentre Felin as it approaches Llangollen on 7 October 2006 while working the 15:20 Carrog–Llangollen service.

Below (Image 198):
46010 (ex D147) approaches Kidderminster (Severn Valley Railway) on 17 May 2019 while working the 11:18 service from Bridgnorth.

Scenes from Llangollen Railway as 46010 (ex D147) …

(Image 199): … eases out of Deeside Halt on 11 March 2006 while working a Carrog–Llangollen service.

(Image 200): … passes Fisherman's Crossing as it approaches Glyndyfrdwy on 6 October 2007 while working the 13:30 Carrog–Llangollen service.

(Image 201): ... powers out of Llangollen on 6 October 6 2007 while working the 12:35 service to Carrog.

Scenes of 46035 (ex D172) **Ixion** *at work on the East Lancashire Railway on 12 July 2001 as it …*

Opposite above (Image 202): ... arrives at Ramsbottom while working the 10:14 Bury Bolton St–Ramsbottom shuttle service. Note the locomotive's reversion to original lined green livery and fleet number as D172, albeit with Ixion nameplate and the final nose-mounted central 4-character headcode box.

Opposite below (Image 203): ... stables in the Ramsbottom run-round siding with the shuttle trainset.

Right (Image 204): 46035 (ex D172) enters Ramsbottom (East Lancashire Railway) on 16 March 2002 while working the 15:00 Bury Bolton St–Rawtenstall service.

Below (Image 205): 46035 (ex D172) poses as a display item at Crewe Works during an Open Day event on 31 May 2003.

Scenes from the East Lancashire Railway as 46035 (ex D172) …

(Image 206): *… arrives at Irwell Vale on 12 July 2001 while working the 13:00 Bury Bolton St–Rawtenstall service then …*

(Image 207): *… awaits the signal to depart for Rawtenstall.*

(Image 208): ... forms the rear of the 12:25 Rawtenstall–Bury Bolton St service powered by Class 45/1 45135 3rd Carabinier on 19 January 2003 as the train departs from Irwell Vale.

(Image 209): 46045 (ex D182) carries its original fleet number as it approaches Bewdley (Severn Valley Railway) on 21 May 2016 while working the 09:55 Kidderminster–Bridgnorth service.

(Image 210): 46045 (ex D182) stables in Swanwick yard (Midland Railway Centre) on 5 August 1997 while awaiting an overhaul and repaint.

SECTION 4
HEADCODE BOXES

The Peak family encompassed the full range of headcode box designs ranging from the corridor connection and nose-mounted discs of the Class 44 locomotives to the nose-mounted central 4-character headcode box of the final Class 46 locomotives. The original build (D1-D10) had the corridor connection and discs while the first Derby-build of Class 45 (D11-D15) retained the corridor but replaced the discs with 2 nose-mounted 2-character headcode boxes. The corridor connection was quickly discontinued hence Derby-built D16-D30 / Crewe-built D68-D107 were built with the pair of nose-mounted 2-character headcode boxes. A further redesign saw Derby-built D30-D49 / D138-D173 and Crewe-built D50-D67 / D108-D137 being built with two nose-mounted central 2-character headcode boxes while the final redesign saw Derby-built D174-D193 receive the nose-mounted central 4-character headcode box design. In the 1960s this final design was applied randomly to both Class 45 and 46 locomotives when they received either works attention or accident repairs.

When the headcode boxes were made superfluous in the mid-1960s they were replaced by a pair of nose-mounted marker lights following the fitting of these to 45071 as noted at **Image 37**. In preservation, however, the locomotive owners were faced with the choice of retaining the marker lights or restoring one of the optional headcode box designs, as noted in the following images.

(Image 211): *A comparison of headcode boxes at Barrow Hill on 6 October 2002 sees* **(left)** *Class 46 46035 (ex D172) with the final nose-mounted central 4-character headcode box and* **(right)** *Class 45/0 45060 (ex D100) Sherwood Forester with original pair of nose-mounted 2-character headcode boxes while …*

Above (image 212): ... the view from the other end shows **(left)** 45060 Sherwood Forester *with a pair of nose-mounted marker lights while* **(right)** 46035 *retains the nose-mounted central 4-character headcode box.*

Right (Image 213): *A comparison of headcode boxes at Ramsbottom (East Lancashire Railway) on 16 March 2002 reveals* **(left)** *Class 45/1* 45135 3rd Carabinier *fitted with nose-mounted marker lights as it forms the rear of the 15:00 Rawtenstall–Bury Bolton St service and* **(right)** *Class 46* 46035 (ex D172) *fitted with nose-mounted central 4-character headcode box powering the 15:00 Bury Bolton St–Rawtenstall service.*

HEADCODE BOXES • 113

(Image 214): THE END *46005 (ex D142) was the first class member to be scrapped after being withdrawn from service in December 1977 following storage at Plymouth Laira at the end of the Summer 1977 timetable. The locomotive was moved to Derby Works during January 1978 for scrapping and where the remains were noted awaiting disposal during a visit on 18 March 1978.*